SHARKS

Rebecca Woodbury, Ph.D., M.Ed.

Gravitas Publications Inc.

SHARKS

Illustrations: Janet Moneymaker

Sharks
ISBN 978-1-950415-68-7

Published by Gravitas Publications Inc.
Imprint: Real Science-4-Kids
www.gravitaspublications.com
www.realscience4kids.com

RS4K Photo credits: Cover & Title Pg: Alexander Vasenin, CC BY SA 4.0; Above–NOAA; P.3. Amada44, CC BY SA 3.0; P.5. Zac Wolf, CC BY SA 2.5; P.7. Matthew Field, www.photography.mattfield.com, CC BY SA 3.0; P.11. Ali Abdul Rahman on Unsplash; P.13. Pascal Deynat Odontobase, CC BY SA 3.0; P.15: Top, PublicDomainPictures from Pixabay; Bottom, K2-Kaji from Pixabay; P.17. Nick Hobgood, CC BY SA 3.0; P.19. Sander van der Wel from Netherlands, CC BY SA 2.0; P.21. Elias Levy, CC BY SA 2.0

Sharks are a type of **fish**.
Fish are a kind of animal that
lives and swims in water.

Blacktip shark

There are many different types
of sharks. Some of them are
very small and some are huge.

Look!
A whale shark!

Whale sharks are
the biggest fish in
the sea.

Whale shark

Most sharks live in the salt water of **oceans**. But some sharks live in fresh water in **rivers**.

Leopard shark

A shark's **skeleton** is made of **cartilage** instead of **bones**. Cartilage is lighter than bone, which makes a shark able to float.

Cartilage can bend. This makes it easier for a shark to swing its tail from side to side so it can swim fast.

Our ears are made of cartilage.

So are kids' ears.

A shark's skin is covered
with tooth-like **scales**.
These make shark skin feel
like sandpaper.

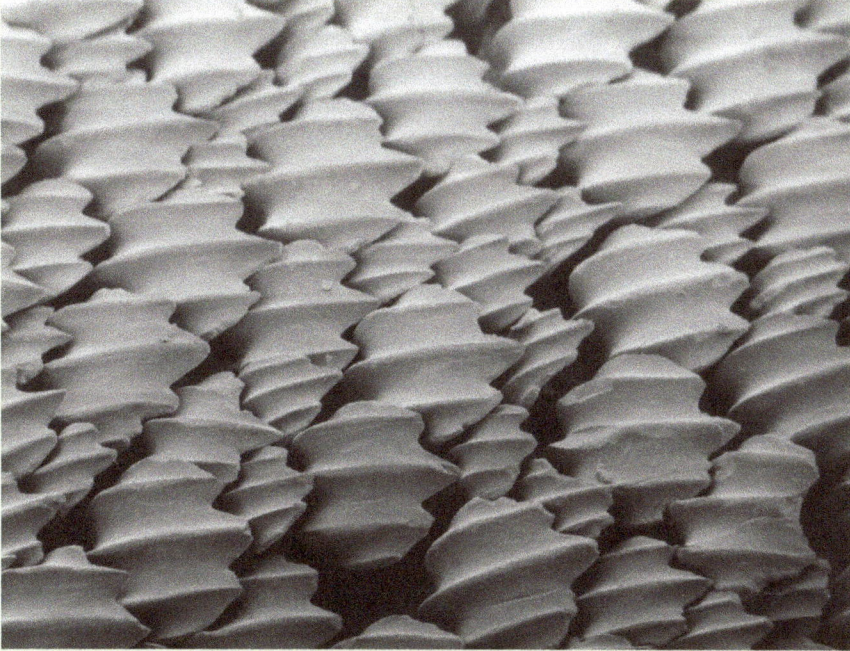

Shark scales enlarged
with a microscope

Sharks have lots of sharp teeth used for catching and eating food. Some sharks have more than one row of teeth.

Did you know that when a shark loses a tooth, a new one will grow back?

Cool!

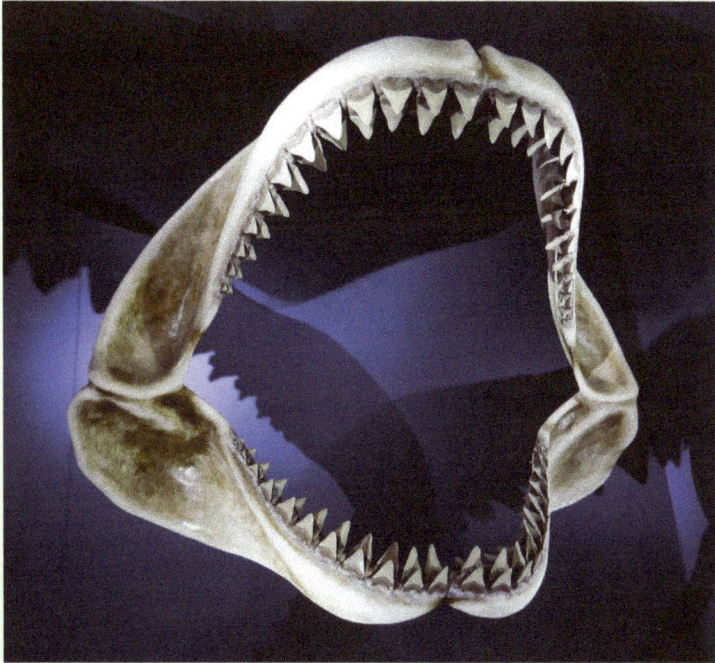

Most sharks eat **fish**, **squid**, and **shrimp**. Some eat **seals** and even other sharks.

Squid

Some sharks lay **eggs** that later hatch into baby sharks. Other types of sharks give birth to live babies.

Sharks have funny-looking eggs.

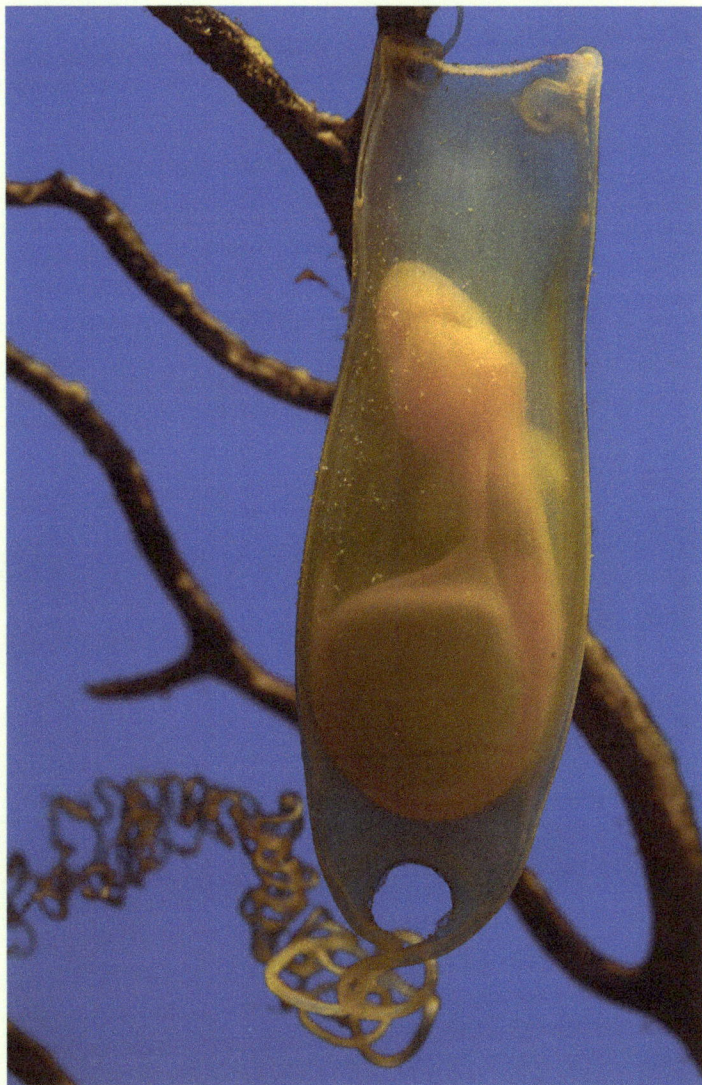

Shark egg

Sharks don't hunt humans. But some types of sharks will bite if people get too close. It is important to stay out of water where sharks swim.

Sharks are very interesting animals. They are an important part of ocean life and need to be protected.

Great white shark

How to say science words

bone   (BOHN)

cartilage   (KAHR-tuh-lij)

egg   (AIG)

fish   (FISH)

leopard   (LEH-puhrd)

ocean   (OH-shuhn)

river   (RIH-vuhr)

scale   (SKAYL)

science   (SIY-uhns)

seal   (SEEL)

shark   (SHAHRK)

shrimp   (SHRIMP)

skeleton   (SKEH-luh-tuhn)

squid   (SKWID)

www.ingramcontent.com/pod-product-compliance
Lightning Source LLC
Chambersburg PA
CBHW040152200326
41520CB00028B/7577